Human Evolutionary Biology

Lab Manual

REVISED PRINTING

Nicole M. Webb | Ryan L. Raaum

Kendall Hunt
publishing company

Cover image © Shutterstock, Inc.
Cover concept Courtesy Nicole Webb.
Header image ©Potapov Alexander/Shutterstock.com

Kendall Hunt
publishing company

www.kendallhunt.com
Send all inquiries to:
4050 Westmark Drive
Dubuque, IA 52004-1840

Copyright © 2017 by Kendall Hunt Publishing Company

ISBN 978-1-5249-6963-9

Published in the United States of America

Contents

Course Introduction

Like all other living things, humans are the result of biological evolution. To develop an understanding of human evolution, you must understand the biological processes that ultimately shape it. Humans are one branch on a vast tree of life encompassing a multitude of forms. Of these forms, the mammals are one diverse group, and within this group is where our own subgroup, the primates, is situated. Not only are our close primate relatives important to understanding human biology, our more distant mammalian relatives can also provide valuable insight into how certain traits evolved in the history of our species. It is easy to forget that we share such affinities with other mammals because over the course of millions of years humans have become both behaviorally and physically unique. Yet, the complexity of our story is only realized when we consider our humble mammalian beginnings as small, nocturnal creatures scurrying about, and then compare it to our current state as a relatively hairless and highly intelligent animal capable of contemplating its own evolutionary trajectory. Beyond our impressive intellectual capabilities, we have also acquired other unique features, including a form of locomotion that is very rare among mammals. What facilitated this critical shift from moving on four limbs to moving only on two? When did humans become so intelligent and begin using the environment in novel ways? What roles did cooperation and culture play in insuring certain evolutionary successes? All of these themes, and many more, will be explored throughout the semester in your human evolutionary biology course.

The lab exercises for this course are designed to support your developing understanding of evolutionary biology and its applications to human evolution, and will reinforce the information you learn in the lectures and readings. The lab manual begins with an overview of basic genetics, and the individual forces responsible for facilitating evolutionary change. As the course progresses, you will apply evolutionary reasoning to the human fossil record and to modern human variation. Instead of considering humans separately from other mammals, you will compare humans to other primates so as to recognize the major adaptive shifts that led to our unique attributes. In turn, you will gain an appreciation for the way in which an animal's physical attributes inform us about aspects of their behavior. You will learn how scientists use evidence to reconstruct evolutionary relationships and determine important features relating to both diet and environment.

Due to the highly integrated and cross-disciplinary nature of physical anthropology, it is inevitable that you will leave this course with knowledge that will inform aspects of your daily life. This is after all, your story. Welcome to the course.

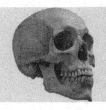

Intended Lab Manual Use

This lab manual is intended to serve as a course guide that will complement the material covered in your human evolutionary biology course. It is designed to reinforce important concepts surrounding studies of human evolutionary biology through providing the opportunity to apply newly acquired information using hands-on lab activities. Each section will include background context, relevant vocabulary terms, and the lab exercises themselves. You should find that the most pertinent information is included in this manual. Additional blank pages have been incorporated at the end of the manual to encourage note-taking as topics are covered in class.

The materials in the manual can, and should be, used during the course as both study guides and as a general reference during the lab activities. This lab manual, however, is not to be used during quizzes or examinations. Be sure to put your name on the lab manual inside cover so that it can be promptly returned to you in the event that it is misplaced.

Given the importance of the lab manual, it is necessary to bring it to class on scheduled lab days and on any other days specified by your instructor. You will note that the pages of the manual are perforated so that you can submit lab exercises upon their completion. It is advised that you do not prematurely remove these labs or their associated content before you are instructed to do so. It is not required of instructors to supply additional lab exercises should you lose or damage the originals.

You will notice in the table of contents that there are eight sections, and consequently eight lab exercises, that will be completed throughout the duration of the semester. It is critical that while performing these lab exercises you adhere to any rules set forth by your course instructor. Important rules have been outlined below but additional ones may apply to your particular course section, so please consult the syllabus and/or your instructor if you have additional questions regarding classroom protocols.

Important Lab Rules

1. **No food or drinks** are permitted on lab days. You may bring a small, covered water bottle and store it in a corner away from any of the lab materials. This is important because liquids and foods can permanently damage the skeletal collections.

2. **Do not use your cell phone** during the lab activities. It is distracting and all of the information you need will be in your class notes or in this lab manual.

3. **Do not pick up skeletal material, or casts, with one hand or by delicate structures** like the eye orbits or the zygomatic arches.

4. **Only work with skeletal material over the provided mats** to avoid the potential for dropping them.

5. **Do not point to the skeletal material/casts with your pen or pencil.** You can turn it around and use the eraser end of your pencil should you need to point to a particular structure.

6. **Avoid carrying or moving the material** unless absolutely necessary and make sure you get permission from the course instructor before doing so.

7. **Do not mix up individual specimens or lab station labels** as it makes it difficult for your peers to study if the material is no longer properly organized.

8. **In the event material is damaged, report it immediately.** Accidents do happen and to minimize the extent of damage the course instructor should be made aware of the situation immediately.

9. **Do not, under any circumstances, remove lab materials from the classroom.** We are very fortunate to have the opportunity to utilize these materials and skeletal collections so please be respectful when working with them to insure their future use.

The Forces of Evolution

Background

Evolution explains how organisms change and diversify over time. Many of the differences in appearance (**phenotype**) between species and between individuals within the same species are, at least in part, the result of differences in their genomes (**genotype**). That is, individuals or species with different phenotypic **traits** have some different genetic variants (**alleles**). Because it is the genome that is directly passed from one generation to the next, the modern definition of evolution is genetic. Evolution occurs when there is a change in the total genetic composition of a population over time. This definition is usually expressed as a change in the **frequency** (proportion) of different alleles: evolution occurs when there is a change to the **allele frequencies** in a population from one generation to the next. The four forces that cause changes in allele frequencies are **natural selection, mutation, genetic drift,** and **gene flow**. These are often collectively referred to as the forces of evolution.

In Lab Exercise 1, you will directly observe how two of the forces of evolution, natural selection and genetic drift, can alter trait and allele frequencies from one generation to the next. We will use beads to represent organisms and the color of the bead will represent the trait of interest. For the purposes of this exercise, we will make the simplifying assumption that there is a one-to-one relationship between the observable trait (bead color) and the underlying genetic cause (allele)—the allele frequencies and trait frequencies are assumed to be exactly equal.

Natural selection occurs when one variant of a trait has an advantage over others, whereas genetic drift is the outcome of chance events. If one trait variant, here bead color, has an advantage, it will become more common in the **gene pool** over time because the individuals that have that trait will survive and reproduce at higher frequencies. These traits that offer an advantage are called **adaptations** because the reason for their success is that they are better suited to the environment. An advantageous or adaptive trait is one that gives an organism a better ability to find food, find a mate, keep safe from predators, or anything else that ultimately leads to continued survival and reproduction. In the natural selection experiment here, different bead colors offer protection against predators through camouflage.

Genetic drift is the result of the inherent randomness in many life events. If individuals have different traits, but none of those traits offer an advantage, then it is random chance that determines which of them have the greatest reproductive success and contributes more to the next generation. Even when some trait variants have an advantage, some individuals with advantageous traits may die early or otherwise fail to reproduce. Similarly, some individuals with less desirable traits may get lucky and have substantial reproductive success. Thus, increases or decreases in the **frequency** (proportion) of specific traits and alleles in a population through genetic drift are, by definition, not related to the features of the organism.

The size of the population itself determines how large the effects of genetic drift are from one generation to the next. In small populations, the loss of just a few individuals can have a large effect on trait and allele frequencies. For example, if there are only 10 individuals in a population and 2 of them are different from the others in some way—say 2 gray squirrels and 8 black—then it is not difficult to imagine that the 2 gray squirrels could lose out: one is electrocuted on a power line and the other is run over by a car. That would be a big change in just one generation, from 20% gray squirrels to none at all. Genetic drift occurs in large populations too, but the size of the effect from one generation to the next is smaller. If there are 200 gray squirrels and 800 black squirrels, it is hard to imagine how all 200 of the gray ones could randomly lose out, so a change of 20% seems very unlikely. But it seems possible that 10 of the gray squirrels might fall victim to random accidents, and that would result in a 0.5% change in the frequency of gray squirrels. This is still genetic drift, just a smaller effect. Genetic drift happens in all populations, but the effect is larger in small populations and smaller in large populations.

While you will not directly examine the other forces of evolution in Lab Exercise 1, mutation and gene flow can also change the frequency of alleles in a population from one generation to the next. Mutations are changes to an organism's DNA sequence caused by copying errors, radiation, chemical damage, and other factors. Changes to a DNA sequence often result in a new allele. Mutations can and do happen in every cell in the body, but only those that happen in the gametes (sperm and eggs) will be **heritable**, allowing them to be passed on to offspring and become established in the population. Mutations occur throughout the genome and may or may not affect the observable phenotype. Those that do affect the phenotype are subject to all of the forces of evolution: further mutation, natural selection, genetic drift, and gene flow. Mutations that have no effect on the phenotype are invisible to natural selection. Mutations are the *only* source of entirely new genetic variation. None of the other forces of evolution introduce entirely new alleles into the world.

Gene flow, or migration, occurs when individuals join or leave a population. Even if the immigrating or emigrating individuals do not carry any unique alleles, their joining or leaving will change the allele frequencies a little bit. For instance, if a population has 20 gray squirrels and 80 black ones (20% gray, 80% black) and a gray one leaves, the trait frequency will change because there are now 19 gray ones and 80 black ones (19.2% gray, 80.8% black). Of course, it is also possible for alleles previously absent in the population to appear because of migration. If we have our same starting population of 20 gray and 80 black squirrels and a brown one moves in, then the composition of the population has changed (now 19.8% gray, 79.2% black, 0.9% brown).

If two populations that are initially connected by gene flow lose that connection—that is, if migration between them ends for some reason—then those two populations will start to become genetically distinct. For instance, perhaps our squirrels live in a large forest and then, because of climate change or logging, the large forest gets broken up into two smaller blocks of forest separated by open ground that the squirrels cannot easily cross. Each separate population will now accumulate their own unique mutations. Then, because genetic drift is random, it is unlikely to lead to identical outcomes in both populations. Natural selection will either maintain existing traits if the environments of the two populations remain the same or select for different traits if the environments change or new advantageous alleles appear through mutation. As time goes on, separated populations will become more and more different from each other. If nothing happens to reestablish gene flow, then **speciation** will occur. That is, there will now be two somewhat different *species* of squirrels rather than just two separate populations. Gene flow is necessary to stop speciation from occurring between populations; if there is no gene flow, speciation is inevitable over any long period of time.

Below is a chart that illustrates how the four forces of evolution affect **genetic variation** within a single population and **genetic differentiation** between two or more populations. You should familiarize yourself with this chart that shows how these forces interact.

Evolutionary Force	Genetic Variation Within a Population	Genetic Differentiation Between Populations
Natural Selection	▼or =	▲ or =
Genetic Drift	▼	▲
Gene Flow (Migration)	▲	▼
Mutation	▲	▲

▲, increase;
▼, decrease;
=, no change.

Natural selection has the most complicated effects on both variation within a population and differentiation between populations, because it is the only force of evolution that results from direct relationships between the traits of an organism and the environment it lives in. Most of the time, natural selection is **stabilizing selection,** maintaining the current traits. If the current trait is already a good fit for the environment, then most changes to that trait will not be as efficient. Imagine opening the hood of a car and making some random change to the engine (a new "mutation"); most of the time the car is not going to run better. In the chart above, the "no change" option for natural selection is the result of stabilizing selection.

If the environment changes, natural selection is often **directional selection.** After environmental change, the current trait may no longer be a good fit to the environment and new variants may be favorable and selected. If gasoline becomes extremely expensive, then the number of gasoline powered cars on the road will decrease and the number of electric cars will increase—a directional shift in the population. Finally, sometimes natural selection is **disruptive selection.** This kind of natural selection occurs when there are different traits within the population that are more or less equally good, but have different strengths and weaknesses in different circumstances. One of the best known examples of disruptive selection is seen in deer mice living in Nebraska. About one quarter of Nebraska is covered by light colored sand dunes that were deposited during the last glaciation. Deer mice living in these Nebraska Sand Hills are light in color, while deer mice living in the adjoining grasslands are dark in color. This difference is due to the divergent selective pressure of predation—lighter colored mice are more camouflaged on the sand hills and darker colored mice are more protected on the grasslands.

Important Terminology

1. **phenotype-** an observable feature or characteristic of an organism.

2. **genotype-** the alleles that an individual has at one or more locations in their genome.

3. **trait-** a distinct form or variant of an observable feature of an organism.

4. **allele-** a distinct sequence at a specific location in the genome; alternate version of a DNA sequence.

5. **locus (plural loci)-** a physical location on a chromosome.

6. **chromosome-** a long strand of DNA with associated proteins located in the cell nucleus. Humans have 46 in their somatic cells and 23 in their gametes (sex cells: sperm and eggs).

7. **diploid-** cells containing two complete sets of chromosomes, one set from each parent. Somatic cells are diploid.

8. **haploid-** cells containing only one set of chromosomes. Sex cells are haploid.

9. **allele frequency-** the proportion of a particular allele in a population.

10. **natural selection-** the process that occurs when individuals with one trait have a reproductive advantage over individuals with other traits.

11. **mutation-** any change in a DNA sequence, which often introduces a new allele.

12. **gene pool-** all of the alleles present within an interbreeding population.

13. **adaptation-** a trait that is present because of natural selection; a trait that does or has offered a reproductive advantage to individuals that have it.

14. **heritable-** traits able to be passed on from parents to their offspring.

15. **speciation-** occurs when an ancestral species splits into two descendent species.

16. **genetic variation-** genetic differences between individuals belonging to the same population or species.

17. **genetic differentiation-** genetic differences between individuals belonging to different populations or species.

18. **stabilizing selection-** occurs when natural selection favors the currently existing average phenotype while selecting against extreme versions of the traits; typically reduces genetic variation.

19. **directional selection-** occurs when natural selection favors one extreme of the trait distribution, causing a shift of the distribution toward the favored phenotype.

20. **disruptive selection-** occurs when natural selection favors both extremes over the intermediate traits; typically results in a bimodal distribution.

21. **heterozygous advantage-** when the heterozygous condition is favored over homozygotes because it provides a higher relative fitness.

22. **homologous chromosomes-** chromosomes containing the same genes, though they may have different alleles, that pair during meiosis (the process responsible for sex cell production).

There are four mechanisms that can lead to evolutionary change from one generation to the next: mutation, natural selection, gene flow, and genetic drift. In this laboratory exercise, we will examine the action of **genetic drift** and **natural selection**. The key difference between these two mechanisms, which can be difficult to get a handle on without direct experimentation, is that natural selection privileges some individuals over others on the basis of their biological traits while genetic drift privileges some individuals over others with no regard to their biological traits.

Experimental preliminaries

For these experiments, you will be working in groups of four students. Each group will be provided with:

- A bowl
- A hand towel (either tan or black)
- A container of tan beads
- A container of black beads

Genetic drift experiment

For this experiment, you will need:

- The bowl
- 10 tan beads (to start)
- 10 black beads (to start)

Follow this procedure:

1. Place 10 tan beads and 10 black beads in the bowl and mix them together.

2. Have one group member reach into the bowl **without looking** and select **10** beads. These 10 beads have been eaten by predators and are no longer in the population.

3. Now look at the remaining beads in the bowl—these are the survivors who will reproduce to create the next generation.

4. Record the **frequency** of the tan beads among the 10 remaining in the bowl in the first generation column for "Round 1" in the data table provided. To calculate the frequency, count up the number of tan beads remaining and divide by the total number of beads. For example, if 4 tan beads remain out of 10, then the frequency is 4/10 = 0.4

5. The beads remaining in the bowl represent the individuals from the population who are allowed to reproduce. Each will contribute two "offspring" to the next generation. For every one of the 10 beads remaining in the bowl, restock the bowl with an additional bead of the same color. Following the same example as above, if 4 tan and 6 black beads remain, restock the bowl an additional 4 tan beads (8 tan beads in the bowl) and an additional 6 black beads (12 black beads in the bowl).

6. Repeat steps 2–4 nine more times, simulating nine more generations of random sampling (genetic drift). If at any point there are only tan or only black beads remain, enter 1 (all tan beads) or 0 (all black beads) in the data table and continue to step 7.

7. Repeat the steps 1–6 one more time, entering your results in the Round 2 line of the data table.

Natural selection experiment

For this experiment, you will need:

- A tan hand towel
- A black hand towel
- 10 tan beads (to start)
- 10 black beads (to start)

Follow this procedure:

1. Spread the **tan** towel flat on the surface of a table.
2. Select one group member to be the "predator" and have them turn away from the towel or cover their eyes so they cannot see the towel.
3. While the "predator" is not looking, the remaining group members will randomly place the 20 beads from the current generation across the surface of the towel.
4. Once all the beads have been placed, the "predator" should turn around or open his/her eyes and, one at a time, pick 10 beads from the towel. The "predator" should attempt to select 10 beads fairly quickly, not taking time to study the towel and should attempt to focus only on the bead that he or she is selecting at any given time.
5. There should be 10 beads remaining on the towel. Count up the number of tan beads among those remaining and calculate the frequency of tan beads. For example, if there are 4 tan beads remaining, the frequency of tan beads is 0.4 (4 tan beads divided by 10 total beads), and enter this value into the first generation column in the data table.
6. The 10 beads selected by the "predator" have been removed from the breeding pool. The remaining 10 beads on the towels will be the parents of the next generation. Pick up all the beads from the towel. Each bead remaining on the towel will contribute 2 offspring toward the next generation. So, continuing our example, if there are 4 tan and 6 black beads remaining on the towel, create a new generation with 8 (2 × 4) tan beads and 12 (2 × 6) black beads.
7. Repeat steps 2–6 nine more times, taking turns as the predator. If at any point there are only tan or only black beads remain, enter 1 (all tan beads) or 0 (all black beads) in the data table and continue to step 8.
8. Repeat steps 1–7 using the **black** towel.

Results and Discussion

After completing the experiments, answer the questions on the backside of the data table sheet. You may submit your answers individually or as a group (or two or three subsets of your group).

Name: _____

Data Table for Genetic Drift Experiment

	Frequency of tan beads (alleles)										
Generation	0	1	2	3	4	5	6	7	8	9	10
Round 1	0.5										
Round 2	0.5										

Data Table for Natural Selection Experiment

	Frequency of tan beads (alleles)										
Generation	0	1	2	3	4	5	6	7	8	9	10
Tan towel	0.5										
Black towel	0.5										

After tabulating your data, answer the following questions.

Questions

1. Compare the results from the genetic drift experiments to the natural selection experiments. Are there any clear differences in the results? If so, what is the reason for the differences?

2. From these experiments, how might genetic drift and natural selection affect natural populations of organisms differently? Is the outcome of genetic drift or natural selection more predictable? How might these different mechanisms of evolutionary change relate to functional traits of organisms?

Stratigraphy and Biostratigraphy

Background

Stratigraphy

Stratigraphy is the study of the sequential layering of sedimentary deposits. The origin, physical characteristics, and spatial relationships of stratified rocks are studied to understand the history of the events documented in the strata. Stratigraphic analysis is based on three core principles.

1. The **principle of superposition** states that, in an undisturbed sequence, each stratum (layer) is younger than the one beneath.

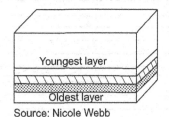

Source: Nicole Webb

2. The **principle of original horizontality** states that strata are horizontal or nearly so when they are deposited.

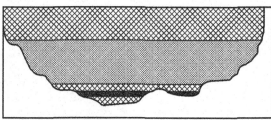

Source: Nicole Webb

3. The **principle of original lateral continuity** states that all parts of a stratum, however disrupted by later activity, were once joined in a single layer.

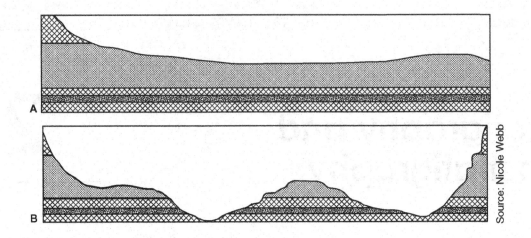

Gaps in the stratigraphic record that result from nondeposition or erosion are called **unconformities**. There are several different types of unconformities and they result when there are episodes of erosion that affect the stratigraphy. When this occurs it makes it difficult to discern the original positions of the individual stratigraphic layers. Take a look at the example provided below. It is a specific type of unconformity called disconformity. This type of interruption in the sediments caused by erosion generates an uneven boundary between parallel layers.

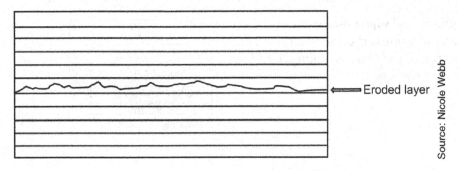

Correlation

Geologists can draw stratigraphic sections for several outcrops in an area and then match beds from one section to another. This is called **correlation**. The sections being correlated are typically within the same general region. Basically, a correlation is a hypothesis that units in two separated sequences are the same. Clearly, the more unique characteristics that two sections share, the greater the probability that the correlation is correct.

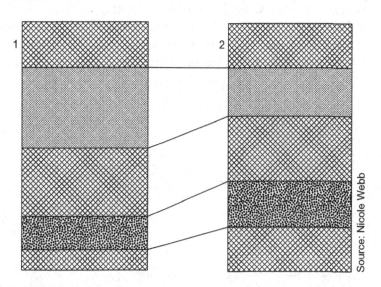

Source: Nicole Webb

When stratigraphic sections are taken from increasingly distant regions, it becomes harder and harder to correlate the layers on the basis of the features of the sediments. For instance, examine the stratigraphic sections below. In these sections, there is only one layer that appears to be equivalent across the sections: bed B in section 3 and bed Y in section 4.

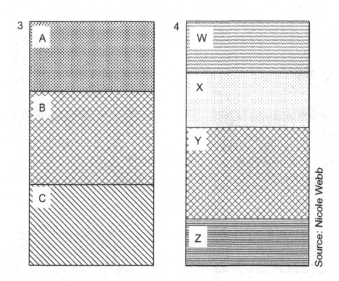

Source: Nicole Webb

Faunal correlation

This is where **biostratigraphy**, the use of fossils in stratigraphic correlation, comes in handy. Representations of some of the fossils present in these sections have been added below. Notice the presence of identical fossil species in layers A and W. Even though the features of the sediments are not the same, they contain the same fossils, suggesting that they were laid down at around the same time. Similarly, the same fossils are present in layers B, X, and Y and in layers C and Z.

To indicate the faunal correlations, we draw lines connecting the layers below the first appearance of the fossil species and above the last appearance (remember, lower layers are older and higher layers are younger). However, we don't draw lines at the very top or very bottom of the sequence, because we don't know if the fossils near the top continue to occur or if the fossils on the bottom would be found even earlier.

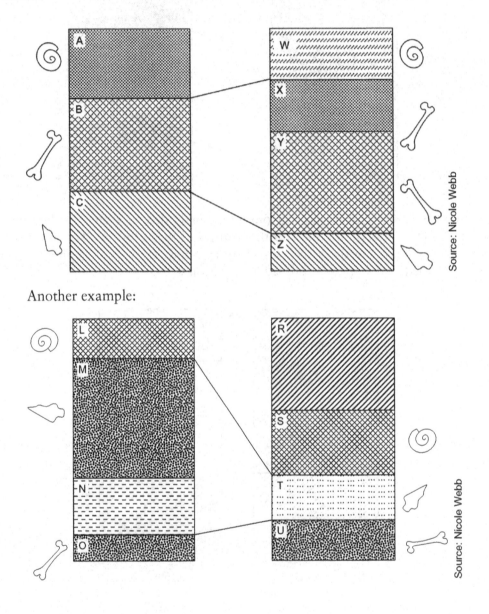

Source: Nicole Webb

Another example:

Source: Nicole Webb

And something a bit more complicated:

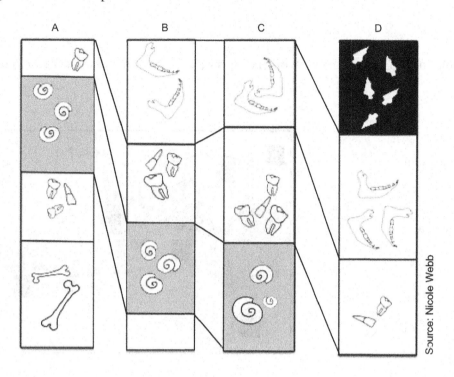

Source: Nicole Webb

Name: _____

A stratigraphic map is shown below with six layers (A–G) and nine artifacts recovered within those layers (1–9).

The artifacts are:

1. Shell necklace

2. Ceramic shards

3. Arrowheads

4. Clay pottery

5. Fish bones

6. Stone Scraper

7. Beads

8. Hominin skull

9. Mastodon remains

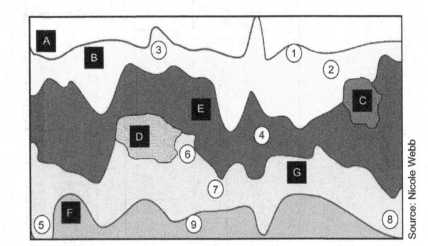

Source: Nicole Webb

1. Which is older, the **arrowhead** or the **clay pottery**?

2. Is **layer G** younger or older than **layer F**?

3. Were there humans living here when **layer B** was being formed? Support your answer.

4. Name an animal that lived here when **layer F** was being formed.

5. **Layer C** is a garbage pit. What layers were being formed when it was dug? What artifacts might have belonged to the person who dug it?

6. Which is older: the **clay pottery** or the **stone scraper**?

7. Which was deposited more recently? The **fish bones** or **layer F**?

8. Correlate the geologic units in the stratigraphic sections below. Draw solid lines connecting all equivalent layers between A and B and between B and C. Do not directly correlate layers A and C. Note that lines cannot cross (think about what this would mean geologically).

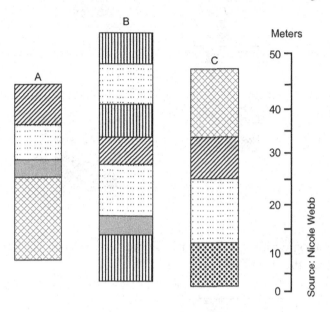

9. Correlate the geologic units in the stratigraphic sections below. Draw solid lines connecting all equivalent layers between A and B. Again, lines can never cross each other.

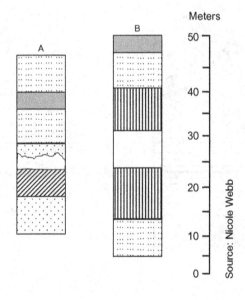

10. Which column (A or B in the stratigraphic sections from question 9 above) contains an unconformity?

11. Correlate the stratigraphic sections below with the aid of the fossils. Draw lines connecting the strata between the profiles.

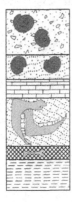 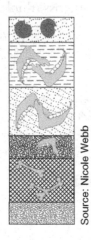

12. Correlate the geologic units in the stratigraphic sections below. Again, draw lines connecting the strata between the profiles.

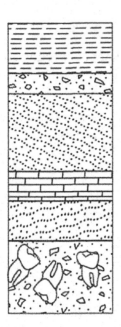 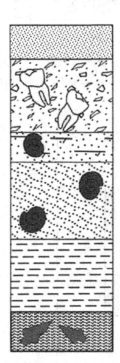 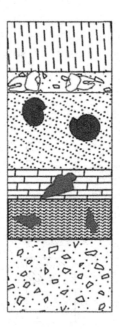 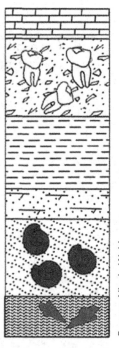

Human Skeletal Biology

Background

The skeleton can be divided into two major divisions:

The **axial skeleton**: the skull, backbone, and ribcage

The **appendicular skeleton**: skeletal elements dealing with limbs: arms, legs, shoulder blades (scapula), and pelvis. Each of these can be further subdivided into individual bones.

Axial skeleton

Skull: The skull consists of the **cranium** (the braincase and face) and the **mandible** (lower jaw). The cranium is, in fact, made up of a number of different bones, which are fused together. The lines where they are fused are called **sutures.** The larger cranial bones are the:

Maxilla: the bones holding the upper

Frontal: forehead area; just above the orbits

Zygomatics: cheek bones

Parietals: behind the frontal, on top (large)

Temporals: side of skull—has the ear opening

Occipital: back and bottom of the skull, has the foramen magnum

Sphenoid: makes up part of orbits and base of skull.

There are a number of other small bones as well (the ethmoid, lacrimal, nasal, vomer and palatine bones, and the three ear ossicles).

Vertebral Column: consists of 7 cervical, 12 thoracic, and 5 lumbar vertebrae.

Ribs: connect to the vertebrae dorsally. Most connect to the sternum ventrally, but some do not.

Sternum: the breastbone. Ventral connection point for all the upper ribs.

Appendicular skeleton

The appendicular skeleton is made up of a number of bones:

Clavicle: the collar bone. Connects the arm to the body.

Scapula: the shoulder blade. It articulates (meets) with the single bone of the upper arm (the humerus) by a *mobile ball and socket joint*, which allows for motion in all directions.

Humerus: the upper arm bone.

Radius and **ulna:** the forearm bones; with your palm facing up, the radius is lateral (the thumb side) and the ulna is medial (the pinky side). Feel these bones on your own arm.

Hand:

Carpals: the wrist bones, consists of eight separate bones.

Metacarpals: the five rod-like bones, which connect the carpals to the phalanges.

Phalanges: the "finger bones". These articulate with each of the metacarpals—there are three phalanges for each finger, except the thumb, which only has two.

Pelvis: This is made up of the:

a) **Innominate** or **os coxa:** the major bone of the pelvis. There is a left and right innominate. Note that the hip socket is always on the lateral side. Each innominate is comprised of three fused bones:

- -ilium: largest bone, covered by hip muscles responsible for moving the hip joint; looks like a wide blade.
- -ischium: posterior to ilium—the bone you sit on.
- -pubis: lies anterior to other two bones; where each ½ joins at the middle.

b) **Sacrum:** connected to the lumbar vertebrae; forms posterior part of pelvis.

Femur: single bone of thigh; it has a round head that articulates with the pelvis, and distal *condyles* which articulate with the *tibia* to form the knee joint.

Patella: The knee cap. It is attached to several muscles of the knee. Thus, the knee is composed of articulations between the femur, tibia, and patella.

Tibia: The larger of the two bones of the lower leg. Part of the knee joint proximally; distally it articulates with the ankle.

Fibula: The lateral and smaller of the two bones of the lower leg. Comparatively slender, it articulates with the tibia above and below.

Foot: This is made up of:

Tarsals: the ankle bones, consists of seven separate bones.

Metatarsals: five rod-like bones that connect the tarsals to the phalanges.

Phalanges: three for each toe, except the big toe, where there are two.

Basic skeletal terminology

Articulate/articulation/joint: the place where two bones are joined

Proximal: for limb bones, towards the end of the bone closer to the articulation with the torso.

Distal: for limb bones, towards the end of the bone farther from the articulation with the torso

Medial: toward the midline of the body

Lateral: away from the midline of the body

Ventral: the stomach side of the body

Dorsal: the back side of the body

Superior: to the top of the body
Inferior: to the bottom of the body
Anterior: toward the front
Posterior: toward the back
Prognathism: the amount to which the face of an animal "sticks out"

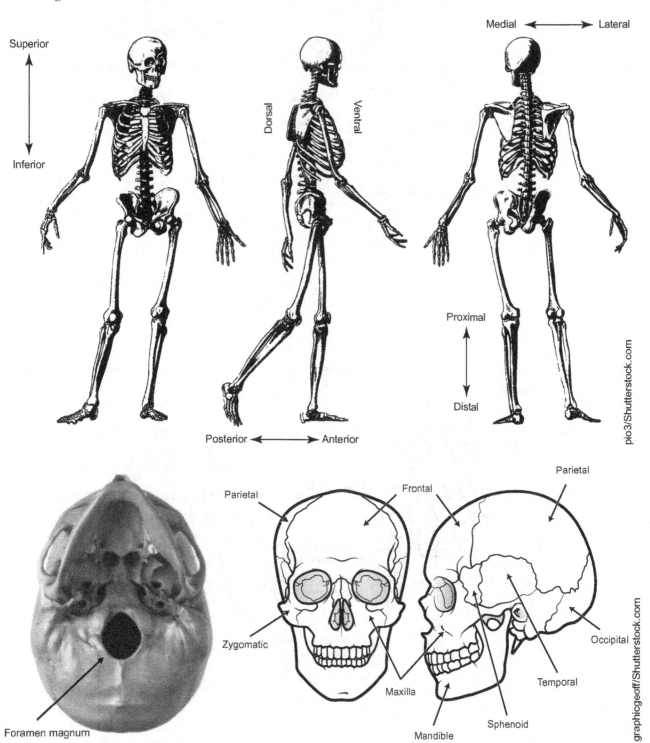

Superior

Inferior

Dorsal

Ventral

Medial ⟵⟶ Lateral

Proximal

Distal

Posterior ⟵⟶ Anterior

pio3/Shutterstock.com

Sebastian Kaulitzki/Shutterstock.com

Foramen magnum

Parietal

Frontal

Parietal

Zygomatic

Maxilla

Mandible

Sphenoid

Temporal

Occipital

graphicgeoff/Shutterstock.com

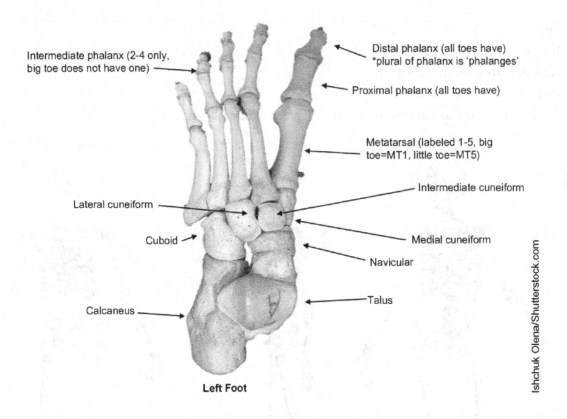

Intermediate phalanx (2-4 only, big toe does not have one)

Distal phalanx (all toes have) *plural of phalanx is 'phalanges'

Proximal phalanx (all toes have)

Metatarsal (labeled 1-5, big toe=MT1, little toe=MT5)

Intermediate cuneiform

Lateral cuneiform

Cuboid

Medial cuneiform

Navicular

Calcaneus

Talus

Ishchuk Olena/Shutterstock.com

Left Foot

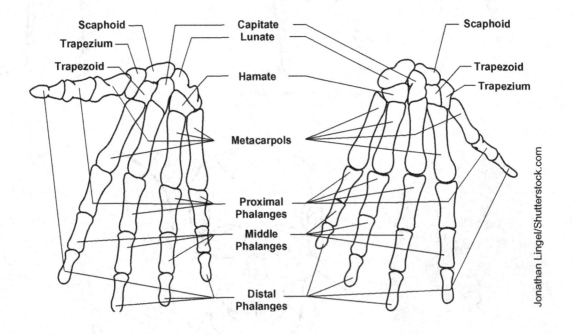

Scaphoid

Trapezium

Trapezoid

Capitate
Lunate

Hamate

Scaphoid

Trapezoid

Trapezium

Metacarpols

Proximal
Phalanges

Middle
Phalanges

Distal
Phalanges

Jonathan Lingel/Shutterstock.com

20

The human skeletal system

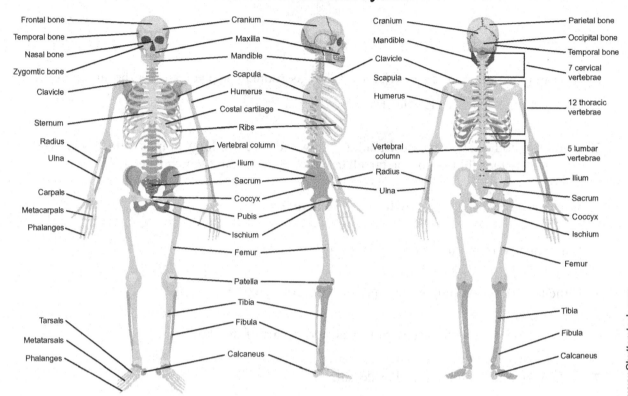

Frontal bone
Temporal bone
Nasal bone
Zygomtic bone
Clavicle
Sternum
Radius
Ulna
Carpals
Metacarpals
Phalanges
Tarsals
Metatarsals
Phalanges

Cranium
Maxilla
Mandible
Scapula
Humerus
Costal cartilage
Ribs
Vertebral column
Ilium
Sacrum
Coccyx
Pubis
Ischium
Femur
Patella
Tibia
Fibula
Calcaneus

Cranium
Mandible
Clavicle
Scapula
Humerus
Vertebral column
Radius
Ulna

Parietal bone
Occipital bone
Temporal bone
7 cervical vertebrae
12 thoracic vertebrae
5 lumbar vertebrae
Ilium
Sacrum
Coccyx
Ischium
Femur
Tibia
Fibula
Calcaneus

www.Shutterstock.com

Name: _____

A. Use the terminology above and the skeletons in the classroom to help you answer the questions below:

1. To what bone does the clavicle attach *medially*?

2. If you feel both sides of your own ankle, you will detect two large bumps. Which bone(s) form these two bumps (hint—not the tarsals)? Which is lateral and which is medial?

3. Which is more distal, the carpals or the humerus?

4. Is a lumbar vertebra part of the axial or appendicular skeleton?

5. Is the foramen magnum on the superior or inferior part of the skull?

6. Are the nasal bones on the anterior or posterior part of the skull?

7. Is the vertebral column ventral or dorsal?

8. How many ribs do you count on one side of the body? Are all of them attached to the sternum?

9. How many bones does the femur articulate with (do not count the patella since it is not a real bone)?

10. Which group of bones are, generally, larger than the other, carpals or tarsals?

B. For each of the following specimens, tell me what the bone is, whether it is part of the appendicular or axial skeleton and, if applicable, what side of the body it comes from (i.e., left or right).

What type of bone is it?	What side of the body?	Axial or appendicular skeleton?
A)		
B)		
C)		
D)		
E)		

Comparative Morphology

Background

Dentition

Different mammals may have different numbers of each kind of tooth (incisor, canine, premolar, molar) in their mouth—but there are some rules. Moving from the front center of the mouth to the back, the teeth are **always** found in the same order: incisors first, followed by *at most* one canine, followed by the premolars, and the molars in the back. Most of the skulls we will see in this course have at least one of each kind of tooth, but some mammalian species have evolutionarily lost some teeth altogether (canines and premolars being the most commonly lost). An informative way to compare the amount of each tooth present across species is by using their dental formulas.

Dental formulas count the specific tooth types found in the mouth if you divide it into four equal quadrants. For example, below is a chart depicting human dentition and the dental formula for humans is 2.1.2.3. This means if you look at half of our upper or lower jaw, you will find 2 incisors, 1 canine, 2 premolars, and 3 molars (some of us have lost our molars due to extraction or, in some cases, they are genetically absent). Our dental formula is the same for the upper and lower jaw so we would write our dental formula as: 2.1.2.3 / 2.1.2.3

Above the line represents the upper jaw (maxilla) dentition and below the line represents the lower jaw (mandibular) dentition. Even though humans have the same number for each, some animals differ which is why both are considered in comparative studies.

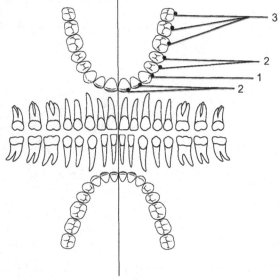

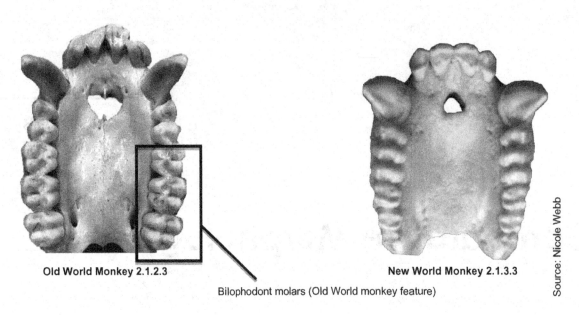

Old World Monkey 2.1.2.3 New World Monkey 2.1.3.3

Bilophodont molars (Old World monkey feature)

Source: Nicole Webb

Dentition can also be informative for inferring diet, especially the molar morphology. Below are illustrations that show features of the molars that are adaptations that allow animals to eat particular foods.

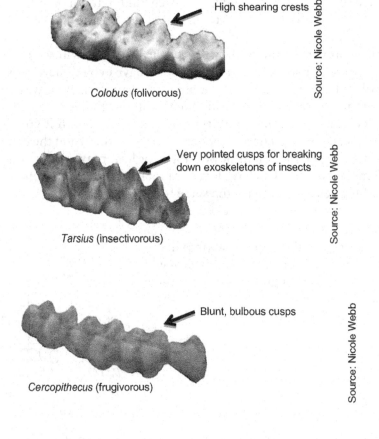

High shearing crests

Colobus (folivorous)

Very pointed cusps for breaking down exoskeletons of insects

Tarsius (insectivorous)

Blunt, bulbous cusps

Cercopithecus (frugivorous)

Source: Nicole Webb

Folivores:
Small incisors

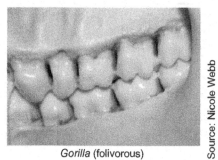

Source: Nicole Webb

Frugivores:
Broad, spatulate incisors

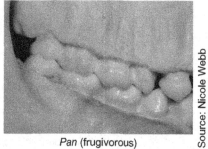

Source: Nicole Webb

Have another look at the molar morphology in a gorilla versus a chimpanzee:

Gorilla (folivorous)

Source: Nicole Webb

Pan (frugivorous)

Source: Nicole Webb

Locomotion

Primates utilize a wide array of locomotor behaviors. They can be subsumed into broad classification types, or modes: **vertical clinging and leaping (VCL)**, **arboreal quadrupedalism**, **terrestrial quadrupedalism**, **suspensory**, and **bipedalism**, but these groups can be broken down further into a variety of subtypes (e.g.,: gibbons are capable of **brachiation**, a type of suspensory behavior that requires hand-over-hand swinging). Some of the larger apes use a form of locomotion called **knuckle-walking** when they move quadrupedally. This list would grow considerably if we were to also include the variety of positional behaviors utilized by living primates.

Locomotor Mode	Features	Taxa	Body Size
1. Arboreal quadrupedalism	• Long tail • Mobile shoulder and elbow joints • Proportional limb lengths (short) • Moderately long hand and foot phalanges	New World monkeys and some Old World monkeys	Small to medium
2. Terrestrial quadrupedalism	• Reduced tail • Limbs are proportional but relatively long • Short, robust, phalanges in the hands and feet • Restricted movement at shoulder and elbow joint for stability	Many Old World species	Medium to large
3. Vertical clinging and leaping	• Large hands • Short, slender forelimb • Long, powerful hindlimbs • Large feet • Long lumbar region	Common in strepsirrhines and tarsiers	Usually small, but some lemurs are medium sized
4. Suspensory	• Long, curved hand phalanges • Mobile wrist, shoulder and hip joint • Short hindlimb	• gibbons (brachiation) • larger-bodied apes and some New World monkeys	Medium to large bodied
5. Knuckle-walking	• Long forelimb • Short hindlimb • Knuckle, wrist, and elbow features that increase stability during flexion • Long, curved hand phalanges	• Chimps and gorillas	Larger ape species

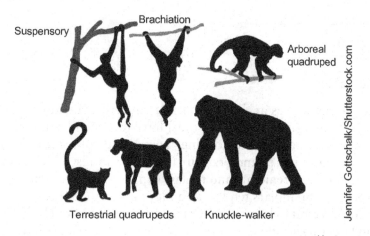

Suspensory

Brachiation

Arboreal quadruped

Terrestrial quadrupeds

Knuckle-walker

Jennifer Gottschalk/Shutterstock.com

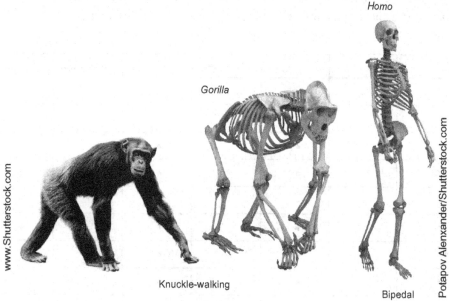

Homo

Gorilla

www.Shutterstock.com

Knuckle-walking

Bipedal

Potapov Alenxander/Shutterstock.com

Relative limb proportions can tell us about locomotor behavior. Suspensory species tend to have much longer forelimbs relative to their hindlimbs. For instance, look at the image above and see how the more arboreal *Pan* compares to a bipedal human. The human has much shorter upper arms, but its use of a bipedal gait is evident by its much longer lower limbs. Both arboreal and terrestrial quadrupedal species will have more proportional limb lengths. To quantify differences in limb proportions among primates the following equation is used to determine their intermembral index:

$$\frac{\text{forelimb length}}{\text{hindlimb length}} \times 100$$

Vertical clingers and leapers (VCLs) have low indices (<90)
Example: lemurs

Suspensory and knuckle-walking species have high indices (>110)
Example: orangutans

Arboreal and terrestrial quadupedal species have intermediate indices (~100).
Example: baboons

Name: _____

STATION 1: The dental formula.

Examine the crania and mandibles at this station and enter their dental formulas in the table below. Some species of strepsirrhines (lemurs and lorisoids) have toothcombs that facilitate grooming and/or certain types of food processing like tree gouging. These toothcombs usually consist of four incisors and two canines, but because the premolars shift forward and resemble canines they are difficult to distinguish. Differentiating the premolars and molars can also be quite challenging—keep in mind that the anterior (most forward) premolars tend to be a bit pointier and more canine-like and the posterior (most rearward) premolars tend to be a bit flatter and more molar-like. The molars themselves typically have more cusps and larger basins than the premolars. Most of the skulls have at least one of each kind of tooth, but *one* of the specimens has evolutionarily lost quite a few teeth.

Species		Incisors		Canine		Premolars		Molars
Ring-Tailed Lemur[1]	Upper		•		•		•	
	Lower		•		•		•	
Howler Monkey	Upper		•		•		•	
	Lower		•		•		•	
Baboon	Upper		•		•		•	
	Lower		•		•		•	
Gibbon	Upper		•		•		•	
	Lower		•		•		•	
Wombat	Upper		•		•		•	
	Lower		•		•		•	

[1]The ring-tailed lemur has a toothcomb! Look at this specimen carefully!

STATION 2: Molar form and diet.

The surface shape of molars and premolars is strongly associated with diet across mammalian species. High, sharp molar cusps and crests are found in species that typically eat tough or fibrous foods, such as meat, leaves, and grasses (species that eat meat tend to have simpler blade-like molars and premolars, while species that eat leaves and grasses and other tough vegetation tend to have more complex patterns of sharp crests). Low, rounded crests with broad basins are found in species that eat more "crushable" or "pulpable" food items like fruits, tubers, and nuts.

Compare the skulls of the bear and the big cat and characterize the features of their molars and premolars in the table below. Not all carnivores are primarily meat-eaters! Many are, but not all. One of these eats mostly meat and one eats mostly fruits, tubers, and other vegetation. Which is which?

	Cusps (Low or High)	Crests (Sharp or Blunt)	Basins (Large or Small)	Basins (Deep or Shallow)	Likely Diet? (Meat or Vegetation)
Bear					
Big cat					

Compare the skulls of the colobine and the cercopithecine and characterize the features of their molars and premolars in the table below. One of them is more of a fruit-eater and one is more of a leaf-eater. Which is which?

	Cusps (Low or High)	Crests (Sharp or Blunt)	Basins (Large or Small)	Basins (Deep or Shallow)	Likely Diet? (More fruit or More leaves)
Cercopithecine					
Colobine					

STATION 3: Locomotion and the primate skeleton.

Look at the feet and hands of the primate skeletons. Long slender fingers and toes (sometimes curved) suggest an arboreal locomotor pattern, while short fingers and toes suggest a terrestrial locomotor pattern. **Describe the length and curvature of the fingers and toes of the two specimens relative to each other.** Be sure to take body size into account! That is, a bigger animal is almost certainly going to have longer fingers than a smaller animal—what we want to know is if either has particularly long/short or curved fingers and toes *for their overall body size.*

A	B

Limb proportions can also tell us about the locomotor habits of primates. Some primates have long forelimbs (arms) and others have longer hindlimbs (legs). Using the ruler, carefully measure a forelimb and a hindlimb of each skeleton.

	A	B
Forelimb length (humerus + radius)		
Hindlimb length (femur + tibia)		

Calculate the intermembral index for each:

	A	B
$\dfrac{\text{forelimb length}}{\text{hindlimb length}} \times 100$		

VCLs have low indices (<90), suspensory and knuckle-walking species have high indices (>110), and quadrupedal species are intermediate (about ~100). According to the intermembral indices, what type of locomotion do you predict for the skeletons you see?

A	B

Primate Features

Background

Primates possess a variety of traits that distinguish them from other nonprimate mammals. These features demonstrate their evolutionary beginnings as arboreal animals that might be described as being squirrel-like in their physical appearance. Their grasping capabilities and binocular, or stereoscopic, vision could have helped them navigate complex arboreal terrains; however, other theories surrounding primate origins also implicate specific foraging strategies or the coevolution of flowering plants as important factors shaping the unique suite of features shared by all primates. This is because if they hunted certain insects nocturnally or exploited flowering plants they might require these same features for sufficient dexterity to catch the prey and to feed at the delicate terminal branches of trees. Further, stereoscopic vision would assist in helping them assess distance to insure they could accurately judge their proximity to their prey, or their ability to walk on certain branch diameters. Despite the various opposing theories, one can appreciate that these features provide numerous advantages; therefore, it could easily be a combination of these ecological pressures that resulted in the distinct assemblage of traits used to distinguish primates.

Eric Isselee/Shutterstock.com

Flattened nails instead of claws.

Primate Characteristics:
Nails instead of claws
Forward-facing eyes
Postorbital closure
Grasping hands and feet
Larger brain relative to overall body size

31

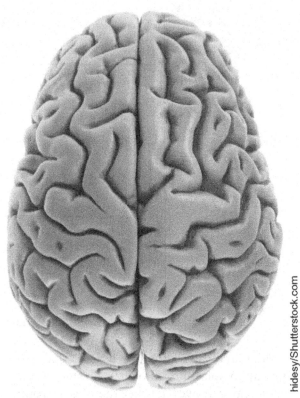

Compared with most other mammals, primates have a larger brain size relative to their overall body size.

Grasping hands and feet.

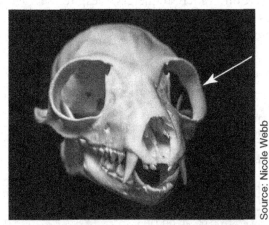

Postorbital closure (bar in strepsirrhines, complete closure in haplorhines).

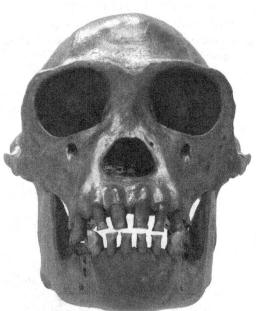

Forward-facing eyes that provide an overlap in the visual fields to provide stereoscopic vision.

Name: _____

STATION 1: Distinguishing Primates from Other Mammals

Looking at the skull, primates are distinguishable from other mammals using a combination of features—no single feature is unique to primates, but the combination is.

Primates have a **closed orbital rim** (either a **postorbital bar** with a ring of bone all around the margin of the eye socket or **postorbital closure** with a completely enclosed eye socket).

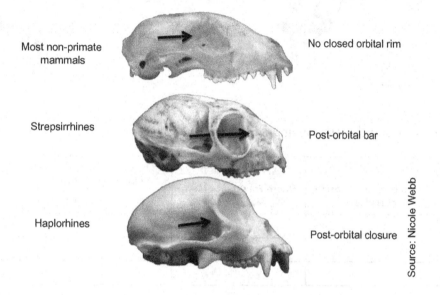

Primates have more **forward-oriented or facing** eyes than many other groups of mammals.

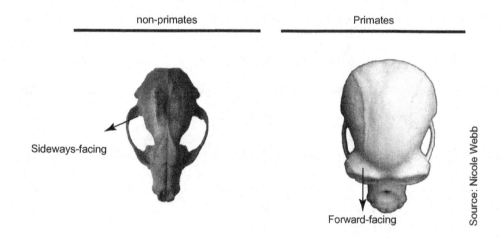

Primates have **relatively large brains for their body size.**

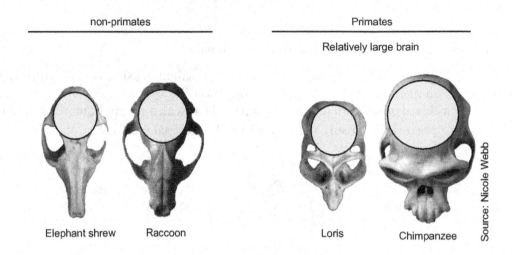

non-primates   Primates

Relatively large brain

Elephant shrew   Raccoon   Loris   Chimpanzee

Source: Nicole Webb

Examine the skulls at this station and enter their characteristics in the table below.

	Orbit (Postorbital Bar, Closure, or Neither)	Eye Orientation (More Forward or More Sideways)	Relative Brain Size (Large or Small)	Primate? (Yes or No)
Skull A				
Skull B				
Skull C				

STATION 2: Differentiating Strepsirrhines and Haplorhines

Within the primates, the two major clades are the strepsirrhines (lemurs, lories, and galagos) and the haplorhines (tarsiers, New World monkeys, Old World monkeys, and apes). There are two features of the skull that can be used to easily differentiate them. First, the **strepsirrhines have a postorbital bar** (ancestral), while the **haplorhines have postorbital closure** (derived). Second, the **strepsirrhines have a tooth comb** in their mandible (derived), while the **haplorhines do not** (ancestral).

Examine the skulls at this station and enter their characteristics in the table below.

	Orbit (Postorbital Bar or Closure)	Tooth Comb? (Yes or No)	Strepsirrhine or Haplorhine?
Skull A			
Skull B			

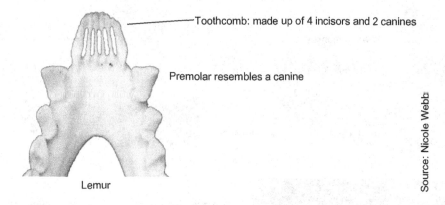

Toothcomb: made up of 4 incisors and 2 canines

Premolar resembles a canine

Lemur

Source: Nicole Webb

STATION 3: Differentiating Platyrrhines and Catarrhines

The two largest clades of haplorhine primates are the platyrrhines (New World monkeys) and the catarrhines (Old World monkeys and apes). There are several features of the skull that are different between these two clades, but we will focus on the easiest to identify: the number of premolars. **Platyrrhines have three premolars** (ancestral) and **catarrhines have two premolars** (derived).

Examine the skulls at this station and enter their characteristics in the table below.

	Number of Premolars (2 or 3)	Platyrrhine or Catarrhine?
Skull A		
Skull B		

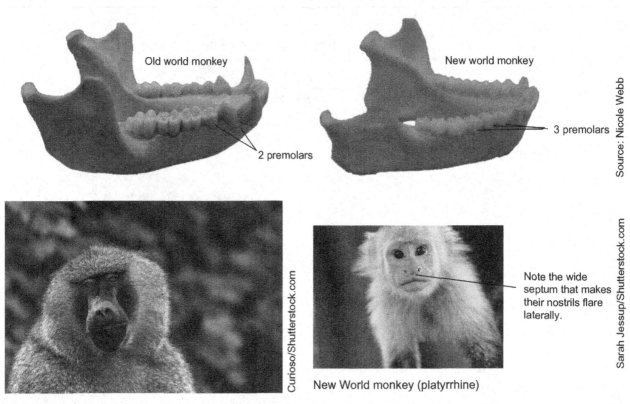

Old world monkey

2 premolars

New world monkey

3 premolars

Source: Nicole Webb

Old World monkey

Curioso/Shutterstock.com

New World monkey (platyrrhine)

Note the wide septum that makes their nostrils flare laterally.

Sarah Jessup/Shutterstock.com

STATION 4: Differentiating Cercopithecoids and Hominoids

Within the catarrhines, there are two clades: the cercopithecoids (Old World monkeys) and the hominoids (apes, including humans). The **Old World monkeys** have a **narrower nose opening** (particularly at the bottom margin), a **narrower palate** (especially at the front), a **somewhat smaller brain** for the same body size, and **bilophodont molars**.

The **apes** have a **broader nose opening** (particularly at the bottom margin), a **broader palate** (especially at the front), a **somewhat larger brain** for the same body size, and **simple molars**.

Examine the skulls at this station and enter their characteristics in the table below.

	Nose (Narrow or Broad)	Palate (Narrow or Broad)	Relative Brain Size (Larger or Smaller)	Teeth (Bilophodont or Simple)	Old World Monkey or Ape?
Skull A					
Skull B					

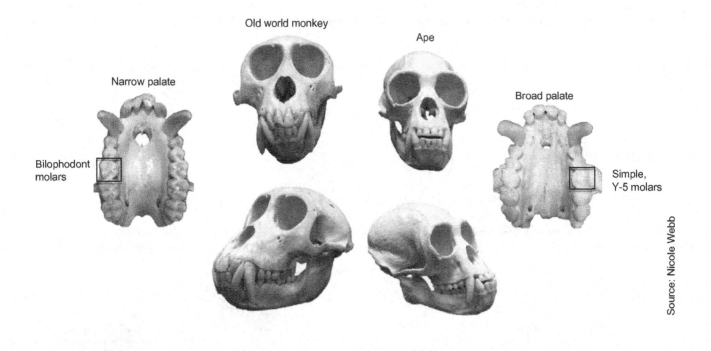

Old world monkey

Ape

Narrow palate

Broad palate

Bilophodont molars

Simple, Y-5 molars

Source: Nicole Webb

STATION 5: Sexual Dimorphism

Many species of primate are **sexually dimorphic**. Sexual dimorphism is a difference in size or shape between males and females. In primates, when there is sexual dimorphism, males typically are larger and have larger canines than females. Among the great apes, all of which are more dimorphic than modern humans, males generally have more sagittal and nuchal cresting, larger brow ridges, and larger canines.

Examine the following skulls of male and female orangutans and gorillas, enter their characteristics in the table below, and answer the question that follows.

	Sagittal Crest (Large, Medium, Small, or None)	Nuchal Crest (Large, Medium, Small, or None)	Brow Ridge Size (Large, Medium, Small, or None)	Canine Size (Larger or Smaller)
Male Orangutan				
Female Orangutan				
Male Gorilla				
Female Gorilla				

For the features you have examined for the table above, orangutan and gorilla males are generally similar as are orangutan and gorilla females. Identify and describe **two** other features of the skull where males and females from the same species are most similar.

1.

2.

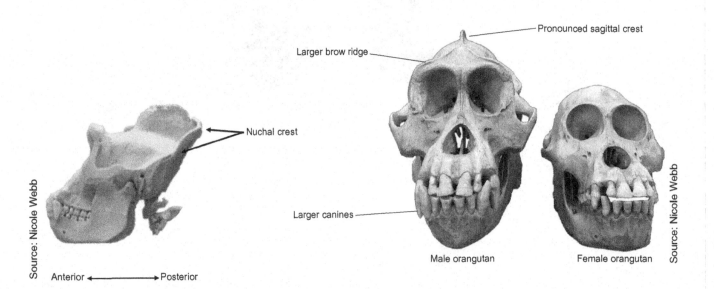

Nuchal crest

Larger brow ridge

Pronounced sagittal crest

Larger canines

Male orangutan

Female orangutan

Anterior ← → Posterior

Source: Nicole Webb

Source: Nicole Webb

Bipedalism

Background

One of the most obvious features of being human is our unusual form of locomotion, bipedalism. Although walking on two limbs is not a unique condition among mammals, our obligate, striding bipedal gait is quite rare. In fact, bipedal movement is only occasionally observed in other primates and is restricted to brief intervals or arboreal contexts.

Humans are *obligate* bipeds because we rely exclusively on bipedalism as our only means of locomotion. Our obligate commitment to bipedalism can be seen in several features of our skeleton. Because the use of a bipedal gait is reflected throughout our skeletons, the transition from arboreal locomotion to the varying degrees of terrestrial bipedalism used by our ancestors is discernable in the fossil record. For this lab activity, you will be assessing the various anatomical indicators scientists use to infer bipedality using skeletal material.

Inferior View of Skull

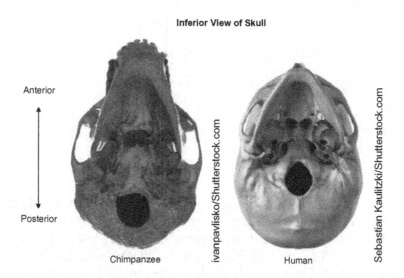

Anterior

Posterior

Chimpanzee

Human

ivanpavlisko/Shutterstock.com

Sebastian Kaulitzki/Shutterstock.com

The **foramen magnum,** which is Latin for "great hole," is an opening found on the bottom of the skull where the spinal column and its associated soft tissue structures pass through. The position of the foramen magnum is helpful for determining the orientation of the skull relative to the spinal column. It provides evidence as to whether the animal is utilizing an upright and potentially bipedal gait. Take a look at the two examples above. The chimpanzee has a foramen magnum that is more posteriorly oriented, meaning it is more toward the back of the skull, when compared to the human which is more anteriorly, or forwardly, placed. This is because the human spinal column is situated under the skull during upright walking, while the chimp is using a quadrupedal stance that requires the spine to enter from the back of the skull. The schematic below illustrates this important difference.

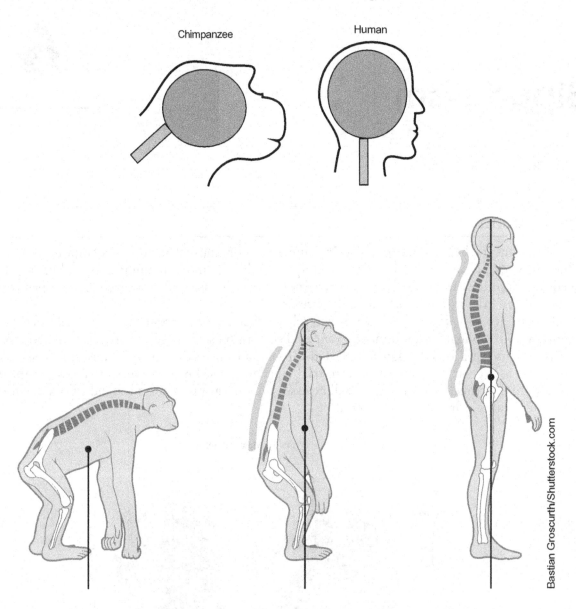

In order to move bipedally, there are several adaptations in the postcranial skeleton that help redirect the center of gravity and successfully allocate body weight down two limbs. The image above clearly illustrates the differences in the location of the center of gravity, represented by the dot, between chimps and humans. Let's take a look at some of the other key differences revealed in the image.

The specific curves of the spine differ in humans because instead of one curve, found in chimpanzees and often described as a "C" shape, humans have two. The upper curve is called the thoracic **kypohosis** and the lower is termed the lumbar **lordosis**. These curves keep the body weight along the center of gravity to efficiently maintain upright posture during bipedal locomotion.

Note the orientation of the pelvis and the extent of observable lateral (side-to-side) flaring. The pelvis is very informative for inferring locomotor behavior because it attaches several of the muscles directly involved in both posture and movement. The wide, bowl-shaped pelvis of humans allows the larger gluteal muscles to attach to it, and these muscles are important for keeping an erect posture and stabilizing the hip during a striding, bipedal gait. The chimpanzee has much smaller gluteal muscles and as you can see from the diagram, the pelvis does not extend laterally, to provide the same degree of muscular reinforcement to the hip joint. You may recall from lecture, the presence of a bony tubercle in bipeds that acts as an attachment site for the *rectus femoris* muscle, a muscle crucial for hip flexion and knee extension during bipedalism, called the Anterior Inferior Iliac Spine (AIIS).

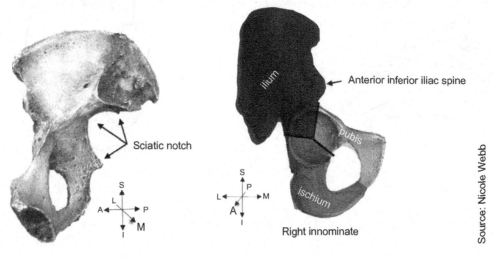

These skeletal attributes, and their proportion/orientation to other bones, also dictate the effectiveness of other muscle groups, namely the hamstrings, in certain loading scenarios. That is why the chimpanzee keeps a bent knee while standing in an erect position. This bent knee is argued by some to make bipedalism less efficient for them compared to humans. Further, the joint size of the lower limbs (hip, knee, and ankle) and the lower spine can be informative, as bipedalism requires larger joint surfaces to accommodate the increased weight sustained on the individual limbs during walking.

In addition, the angle for which the knee joint articulates is different in that humans have what is called a valgus knee angle that serves to redirect the weight of the body medially (toward the body's midline).

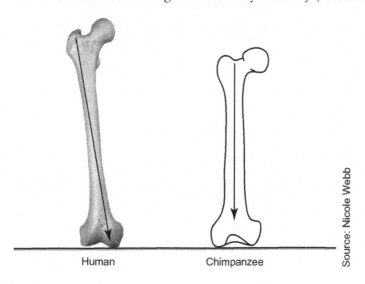

41

The foot is also a highly informative region for studying locomotion. Early in hominin evolution, we see the retention of an opposable big toe, also called a hallux, which is evidence for their arboreal past. As our ancestors began using more bipedalism, the hallux became more aligned with the other toes to eventually form the configuration seen in humans. Bipedal adaptations include the presence of a longitudinal and transverse arch, which act as important shock absorbers throughout bipedal gait.

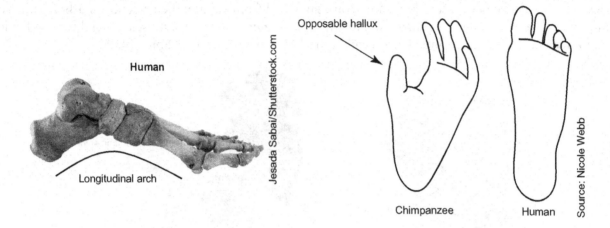

Name: _____

STATION 1: The Foramen Magnum

In bipeds, the foramen magnum is located in a more anterior position on the underside of the skull. Examine the skulls at this station and draw in the position of the foramen magnum on the image in the first column. Then, for the fossil hominins, describe the position of the foramen magnum relative to the chimpanzee (nonbiped), the modern human (biped), or both. For example, "slightly more forward than the chimpanzee," "between the chimpanzee and modern human," and "very close to the modern human position".

	Draw the foramen magnum on the image	Describe the position of the foramen magnum
chimpanzee	Source: Nicole Webb	Near the back of the cranium
Australopithecus	Source: Nicole Webb	
Paranthropus	Source: Nicole Webb	
Small Homo	Source: Nicole Webb	
Large Homo	Source: Nicole Webb	
Modern human	Source: Nicole Webb	Near the middle of the cranium

STATION 2: The Innominate/Os Coxa

The earliest morphological features of bipedalism to occur in the hominin fossil record are the changes in the innominate or os coxa. These features include a **shorter and wider ilium** that is **angled forward,** thereby creating the **sciatic notch.**

	Shape of the Ilium (Short and wide or Long and narrow)	Orientation of the Ilium (Parallel to back or Angled forward)	Sciatic Notch (Present or Absent)
Chimpanzee			
Australopithecus (Sts 14)			
Paranthropus (SK 3155)			
Large *Homo* (KNM-ER 3228)			
Small modern human (Pygmy)			
Large modern human			

STATION 3: The Femur

Morphological features of the femur give important evidence on hominin locomotion. Some of the key features that can be seen in the fossil record are the valgus angle at the knee joint between the femur and the tibia: in bipeds, this joint is **angled**, whereas in nonbipeds, the femoral-tibial joint is not. In modern humans, the **femoral head is relatively large**. In many fossil hominins, the **femoral neck is relatively long**.

	Angle at Knee (Vertical or Angled Inwards)	Relative Size of Femoral Head (Larger or Smaller)	Length of Femoral Neck (Longer or Shorter)
Chimpanzee			
Australopithecus (Sts 34/TM 1513)			
Paranthropus (SK 82)			
Small *Homo* (KNM-ER 1472)			
Modern human			

Brain, Diet, and Tools

Background

Brain size and skull shape

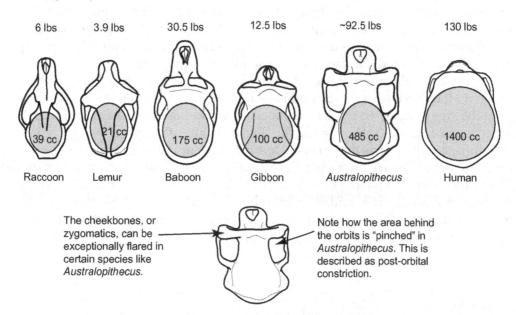

Average Body Size

| 6 lbs | 3.9 lbs | 30.5 lbs | 12.5 lbs | ~92.5 lbs | 130 lbs |

| 39 cc | 21 cc | 175 cc | 100 cc | 485 cc | 1400 cc |

| Raccoon | Lemur | Baboon | Gibbon | *Australopithecus* | Human |

The cheekbones, or zygomatics, can be exceptionally flared in certain species like *Australopithecus.*

Note how the area behind the orbits is "pinched" in *Australopithecus*. This is described as post-orbital constriction.

Primates have large brains relative to their body size when compared to most mammals. This is a trend that becomes more pronounced throughout hominin evolution. In fact, during the last 2 million years our ancestors have substantially increased their brain size. The increase in brain size has been attributed to critical shifts in diet, namely the consumption of high-quality foods like meat, and enhanced cognitive abilities evidenced by their use of stone tools. In addition, there have been other changes to the skull that provide insight into what they were eating. This includes a reduction in **prognathism**, or facial protrusion, and dentition that is much smaller than that seen in the earlier *Australopithecus* and *Paranthropus* species. The image above shows the variation in both skull morphology and the degree of prognathism observed across primates. Note the drastic reduction in facial prognathism seen in the modern human compared to even closely related taxa like *Australopithecus*.

EVOLUTION OF THE SKULL

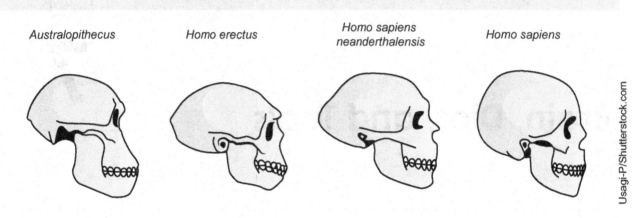

Australopithecus   Homo erectus   Homo sapiens neanderthalensis   Homo sapiens

Usagi-P/Shutterstock.com

Diet

There are four muscle groups involved in mastication (chewing)—the masseter, the temporalis, and the lateral and medial pterygoids. The masseter and temporalis are superficial muscles and in primates their insertion sites mark the cranium and can inform our understanding of their bite force capabilities and their potential dietary preferences. The larger these muscles are, the more force they can generate as the muscles contract and bring the mandible, or jaw, to the cranium. In humans, we have very gracile chewing muscles relative to earlier hominins, this is evidence of a critical dietary shift, perhaps facilitated by the ability to cook our food. We can assess these features comparatively across taxa by looking at several features including the insertion site of the temporalis muscles, which leaves a slight depression on the cranium itself referred to as the **temporal fossa.** Even the positioning of the zygomatics can tell us how large the chewing muscles were and their exact orientations because they pass through, or connect directly to, the zygomatic arches. In addition, the size and shape of the mandible can be informative for inferring bite force capabilities as they also anchor masticatory muscles. And finally, dentition is very useful for reconstructing dietary preferences; specifically, the size and shape of the premolars and the molars. For instance, folivores tend to have larger teeth for their body sizes relative to frugivores, along with the presence of shearing crests and other morphologies that are adaptations for mechanically processing fibrous plant material. Outside of the genus *Homo*, we also see most hominins have molarized premolars that contrast with the reduced premolar morphology noted in anatomically modern humans.

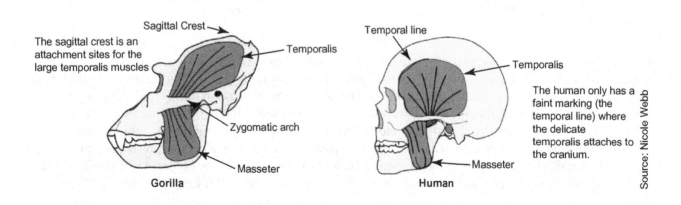

The sagittal crest is an attachment sites for the large temporalis muscles

Sagittal Crest
Temporalis
Zygomatic arch
Masseter

Gorilla

Temporal line
Temporalis
The human only has a faint marking (the temporal line) where the delicate temporalis attaches to the cranium.
Masseter

Human

Source: Nicole Webb

Stone tools

Stone tools (or "lithics") are important for understanding human evolution because they represent a practice that demonstrates our ancestor's capacity to generate tools with a specific purpose mind. This process involves extensive forethought and consequently, speaks to their level of technological innovation and creativity. Although there are a variety of stone tool industries, this lab will focus on only two of them, the Oldowan and the Acheulean. These two industries can be found at the same sites and often even overlap temporally. This has been interpreted as evidence that perhaps different hominins used different industries at these shared localities, or alternatively, that they may have utilized the different tool types depending on the specific task at hand. For instance, one can liken it to the fact we still have pencils despite the invention of the more advanced ballpoint pen.

The Oldowan (2.6–1.7 Ma) is an early Lower Paleolithic industry found in Africa, the Middle East and Europe. Although it is primarily found in association with late *Australopithecus* and small *Homo*, it is named after the Olduvai Gorge site in Tanzania where it has also been found with *Paranthropus* remains. This raises questions regarding who the original manufacturer of these tools actually was, especially in light of a recent discovery of an even earlier tool industry called the Lomekwian that predates *Homo*.

At around 1.7 Ma, a different tool industry, the Acheulean, which is believed by some researchers to be a more refined version of the preceding Oldowan, is usually found in association with early representatives of large *Homo*. This industry has an extended geographical range and an impressive temporal range that extended through much of hominin evolution. The Acheulean is characterized by the production of bifacial handaxes (pictured below). This is different from the tools found in the Oldowan, as Oldowan tools are created by using what is referred to as a hammerstone to pound stone surfaces to produce flakes that are then used for their sharp edges. In contrast, the handaxes characteristic for the Acheulean involve flaking of both sides until they produce cores that are more symmetrical and form tear, or pear-shaped implements. Handaxes can be retouched using other materials, like wood or bone, to carefully chip away additional, small flakes resulting in very detailed edges. The Oldowan and Acheulean can be differentiated by looking at the finished tools, both sides of the Acheulean tools are worked, whereas the Oldowan flakes are only altered on the side where it was detached from the original core.

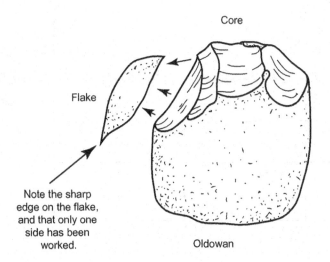

Core

Flake

Note the sharp edge on the flake, and that only one side has been worked.

Oldowan

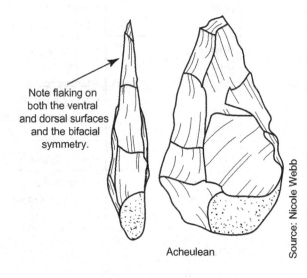

Note flaking on both the ventral and dorsal surfaces and the bifacial symmetry.

Acheulean

Source: Nicole Webb

Important terminology

Brain and cranial morphology

1. **cranial capacity**: a measure of the cubic capacity of the braincase. It can be based off of linear cranial measures that are used in formula or by filling the skull with beads, rice, or some other small element that can be used to infer volume. Often measured in cubic centimeters (cc or cm^3).

2. **absolute brain size**: the actual size of the brain itself.

3. **relative brain size**: the size of the brain compared to the overall body size.

4. **prognathism**: the degree of mid-face projection, or how far the middle region of the face protrudes.

5. **postorbital constriction**: when the area immediately behind the eye orbits tapers in, or appears to be "pinched" relative to the rest of the cranium. Very common in primitive hominins like *Australopithecus*.

Diet

1. **mastication**: the act of chewing to break down food using the teeth.

2. **muscles of mastication**: temporalis, masseter, medial pterygoid, and the lateral pterygoid.

3. **maxilla**: the upper jaw bone.

4. **mandible**: the lower jaw bone.

5. **Types of primate diets**: frugivores (fruit), folivores (leaf/plant material), insectivores (or faunivores), gummivores (gums).

6. **shearing crests**: folivores and insectoivores are equipped with these high crests on their molars that help them break down fibrous plant material.

Diet Type	Features	Example	Body Size
Frugivore (fruit)	• broad incisors • low rounded molar cusps • long small intestine	Orangutan	medium to large
Folivore (leaf and plant material)	• shearing crests • small incisors • large cecum • complex stomach • enlarged large intestine	Gorillas	medium to large
Insectivore (bugs)	• sharp cusps • short, simple gut	Tarsiers	small
Gummivore (gums and tree saps)	• stout incisors • procumbent incisors • long cecum	Galagos	medium to small

Stone tools

1. **Lithic Reduction:** The process of detaching lithic flakes using a variety of materials (e.g., hammerstones, wood, antler, and bone).

2. **Hammerstone:** hard stone cobbles used to remove flakes off of a stone tool.

3. **Core:** what is left of the stone after lithic flakes are removed during lithic reduction.

4. **Flake:** The portion of stone removed via percussion during lithic reduction.

5. **Oldowan tool industry (~2.6–1.2 Ma):** predominately choppers, found in association with early *Homo* and possibly *Paranthropus*.

6. **Acheulean tool industry (~ 1.7–0.2 Ma):** mostly characterized by its distinct handaxes. Typically found in association with *Homo erectus*.

Name: _____

STATION 1: The evolution of face and brain size and shape in hominin evolution.

Examine the crania at this station and characterize them for the characteristics of the face and brain listed in the table below.

	Absolute Brain Size (Rank from smallest [6] to largest [1])	Braincase Shape (Globular or oblong)	Prognathism (Rank from least [6] to most [1])	Size of Face Relative to Brain (Rank from smallest [6] to largest [1])	Position of Face (Front or under)
Chimpanzee					
Australopithecus					
Paranthropus					
Small Homo					
Large Homo					
Modern human					

STATION 2: Bony correlates of chewing musculature on the cranium.

Examine the crania at this station and characterize the bony attributes that are relevant to estimating the size of the chewing musculature.

	Upper Edge of Temporalis (sagittal crest, strong temporal line, or no temporal line)	Postorbital Constriction (Rank from least [5] to most [1])	Size of Temporal Fossa (Rank from smallest [5] to largest [1])	Size of Zygomatic Arch (Rank from smallest [5] to largest [1])	Degree of Zygomatic Flaring (Rank from smallest [5] to largest [1])
Chimpanzee					
Australopithecus					
Paranthropus					
Large Homo					
Modern human					

STATION 3: The mandible.

Examine the mandibles at this station and fill in the table below.

	Ramus Size (Rank from smallest [5] to largest [1])	Molar Size (Rank from smallest [5] to largest [1])	Premolar Size (Rank from smallest [5] to largest [1])	Premolar Shape (More or less molar-like)
Chimpanzee				
Australopithecus				
Paranthropus				
Large Homo				
Modern human				

Follow-up to Stations 2 and 3: Temporalis and Masseter Size
Your observations of the cranium and mandible at Stations 2 and 3 should allow you to make an estimate of the sizes of the temporalis and masseter muscles in these taxa.

First answer the following questions:

1. What features from Stations 2 and 3 are relevant to estimating the size of the temporalis?

2. What features from Stations 2 and 3 are relevant to estimating the size of the masseter?

Now list the groups in the following table by your estimate of the size of their chewing muscles:

	Estimate of Temporalis Size	Estimate of Masseter Size
Smallest		
Largest		

Finally, answer these questions:

3. Taking into account the tooth size and shape information from Station 3, which group is most likely to have had the diet that is most challenging to chew?

4. Which group is most likely to have the easiest diet to chew?

STATION 4: Stone tools.

Different types of tools are often named after the first place at which they were found. For instance, the Oldowan tools you will be looking at today were named after Olduvai Gorge, where they were first discovered. Oldowan tools are typically associated early *Homo*. Compare them to the Acheulean tools, which are typically associated with late *Homo*.

	Oldowan Tools	Acheulean Tools
Overall size of flakes and cores (larger or smaller)?		
Are the **cores** symmetrical (yes or no)?		
Are the **flakes** symmetrical (yes or no)?		
Is there any evidence of **retouching** (smaller chipping to sharpen and shape the edge) on either the cores or flakes (yes or no)?		

5. Which of these tool types would be harder to make?

6. What does your answer to question number 5 imply about the relative cognitive capacities of Oldowan and Acheulean toolmakers?

Modern Human Variation

Background

As discussed in previous sections, natural selection favors certain traits that enable organisms to survive in their environments. These traits are termed **adaptations**, and humans have several adaptive traits that illustrate the types of selective pressures they faced throughout their evolution. Humans adapt to environmental stresses through a combination of both biological and cultural means. For this lab, you will be examining some of the biological adaptations that have shaped human evolution, including those that provide the notable phenotypic variation seen across modern human populations. You will also learn about the distinctions between permanent adaptations, also referred to as genetic adaptations, which do not change throughout an individual's life, and short-term adaptations that temporarily alter the phenotypes of individuals over the course of their lifetime. The following bolded terms will help clarify the different extents of adaptive capabilities, though these distinctions are difficult given the complex interplay of biological, cultural and environmental influences.

Acclimatization: changes that occur over a short period – days to weeks – in an organism to minimize the harmful effects that result when exposed to certain environmental conditions, these may include insufficient nutrient intake, disease and features of the climate. The ability to tan when exposed to ultraviolet rays (UVR) from the sun is one example of acclimatization in humans.

Although one's genetic make-up determines their tanning capabilities, exposure to ultraviolet rays (UVR) is necessary to induce the darkening of the skin. Further, this process is clearly reversible in that tans fade when the UVR is reduced, which is why it is specifically an example of acclimatization. Biologically, the ability to tan is related to the presence of **melanocytes** located in the epidermis layer of skin that release a pigment called **melanin**. Melanins are not the sole biological contributors to skin pigmentation, but they do play a huge role in determining skin color because of their ability to absorb visible light. Of the various forms of melanins, **eumelanin** is the most important type for tanning. Individuals with higher eumelanin concentrations have darker skin color, while the opposite is true for those that have less.

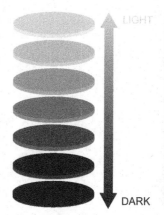

LIGHT

MELANIN

DARK

gritsalak karalak/Shutterstock.com

As you will see in your lab activity, skin color in modern humans produces a clinal distribution that correlates with latitude. Consequently, regions with high concentrations of UVR, individuals tend to have darker skin pigmentation.

Developmental adaptations: changes that occur during development that affect critical growth phases can result in permanent changes to the individual's phenotype. For instance, several of the adaptations required for living in high altitudes are examples developmental adaptations as those that reside in these regions tend to have larger lungs that allow them to increase their oxygen intake efficiency in environments with lower oxygen levels.

Biogeographical Trends

So far several short-term or gradual adaptations have been discussed, but there are also broader trends that relate to biogeography. In fact, most animals that **thermoregulate,** or those that control their body temperature internally, have body plans that are better suited for particular temperatures. Changes to overall body size and relative proportions can aid one's efficiency in regulating internal body temperature. For instance, it is often observed that warm-blooded species living in colder climates generally have larger bodies compared with those residing in warm, tropical environments, a trend referred to as **Bergmann's rule.** This is because they have more volume relative to surface area to generate and maintain metabolic heat. In fact, the more spherical the body is, the more efficiently it can maintain heat. The opposite is true for slender builds, as the surface area is higher, providing more opportunities for heat to dissipate through the skin during circulation. Further, cold-adapted populations tend to have shorter limb proportions for their bodies while those living in hotter climates have longer appendages, a trend called **Allen's rule.** This serves the same adaptive purpose of minimizing surface area in those trying to conserve heat and maximizing it in those concerned with adequately cooling their bodies in warm temperatures. Think of the proportions of a giraffe, the long and slender limb provide much more surface area when compared to a cold-adapted animal like a polar bear with its significantly shorter limbs. The image below shows the relationship between volume and surface area to help conceptualize the adaptive significance of certain body proportions.

Bergmann's Rule = species residing in colder climates have larger body sizes.

Allen's Rule = volume-to-surface area ratios vary for endothermic mammals depending on climate. In colder climates, shorter appendages and more rounded body forms will conserve heat better because there is less surface area for the body heat to escape through. In warm environments, more linear body forms with longer appendages aid in effective heat dissipation.

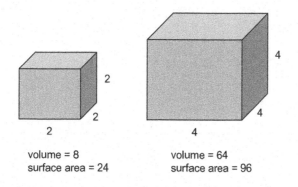

volume = 8
surface area = 24

volume = 64
surface area = 96

Source: Nicole Webb

As mass increases, available surface area decreases proportionately.

Important Terminology

1. **adaptation**: a trait that confers an advantage to the organism and is maintained by means of natural selection.

2. **acclimitization**: The process that allows an organism to respond to changes in the environment such as temperature or altitude.

3. **developmental adaptation**: changes that occur throughout development and result n permanent changes to one's phenotype.

4. **melanocytes**: melanin-producing cells found in the skin

5. **melanin**: the pigments responsible for the observable color differences in human skin, hair and eye color.

6. **eumelanin**: most common type of melanin. Two separate types exists, one that is black and one that is brown.

7. **thermoregulation**: the process that allows the internal body temperature to maintain and regulate a core temperature despite the external temperature.

8. **Allen's Rule**: states that body form is more linear in warm-adapted individuals and more round in those living within cold climates. This is because the former dissipates heat while the latter is more efficient at conserving it because it provides less surface area.

9. **Bergmann's Rule**: states that individuals living in colder climates have relatively larger body size compared to warm dwelling organisms that are smaller. Like Allen's rule, this rule is based on available surface area relative to volume. Long and thin bodies dispel heat more than large and rounded individuals.

Name: _____

Exercise 1: Skin Color

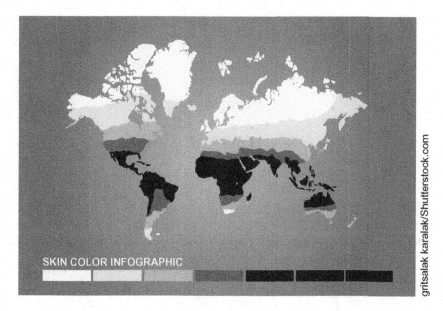

1. When looking at the map of skin color distribution above, what geographical regions have individuals with the highest amount of eumelanin? What do these separate regions all have in common?

2. What are some selective pressures that have shaped the evolution of skin color variation within humans?

3. What are some potential disadvantages of having light skin? Dark skin?

4. Early modern humans in Africa around 200 ka probably had skin of what color?

Name: _____

Exercise 2: Altitude

1. How can humans acclimatize to high altitudes?

2. What benefit does this acclimatization give to an individual visiting higher altitudes?

3. What types of physical changes are noted in people that actually grow up at higher altitudes?

Name: _____

Exercise 3: Bergmann's and Allen's Rules

Examine the figures below and answer the following questions.

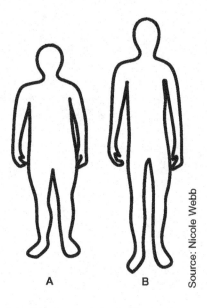

A   B   Source: Nicole Webb

1. Is person A adapted for a cold climate or a hot climate?

2. Which individual has more surface area? Would they be better or worse at conserving heat?

3. What do you notice about their limb proportions?

4. Is this person B adapted for a cold or a hot climate?

5. Describe one trait that helped you make that determination.

6. Is this trait related to Bergmann's rule or Allen's rule?

7. Can you think of other adaptations in humans that allow them to live in warm/ cold climates?

Appendix

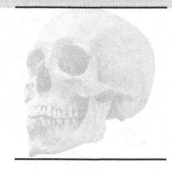

Geochronological scale

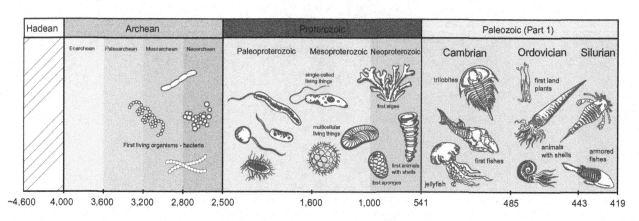

million of years ago

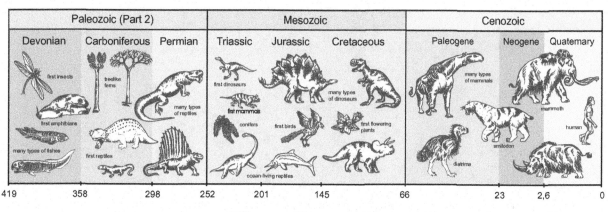

million of years ago

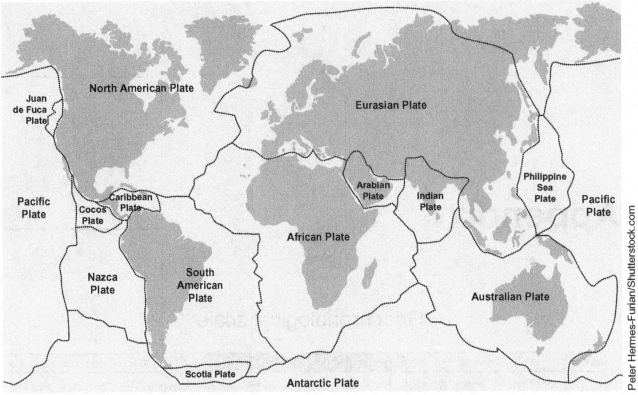

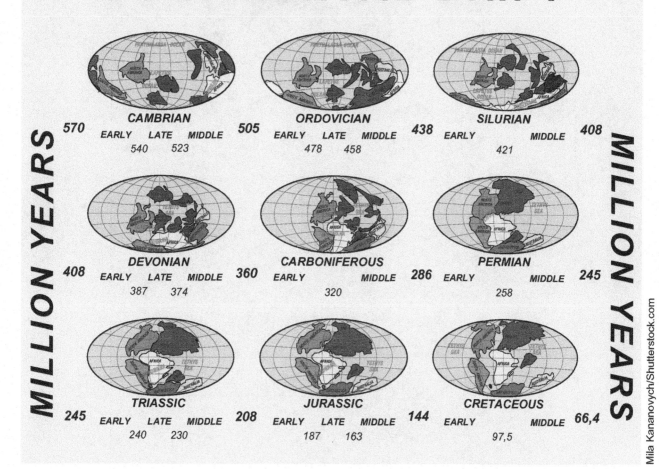

NOTES

NOTES

NOTES

NOTES

NOTES

NOTES

NOTES

NOTES